First edition
2024

Welcome to the Microbial Comedy Club, the tiniest hot spot on the petri dish map, where the world's most charismatic microbes meet and share a laugh.

Here, every microorganism gets a chance to step up to the mic and introduce their unique traits to an audience of their peers.

In this club, bacteria crack jokes about their cell walls, viruses poke fun at their capsids, and fungi tell tales of their mycelial adventures.

It's a place where ribosomes laugh out loud, and the only things infectious are the jokes

Each night, a different microbe takes center stage to showcase its unique characteristics. From the rod-shaped comedians with their curvy sense of humor to the spherical storytellers bursting with tales from the bloodstream, everyone has a story that's too good to stay contained.

This is the Microbial Comedy Club – where every microbe has its moment, every laugh is louder than the last, and every character is contagious in the best possible way!

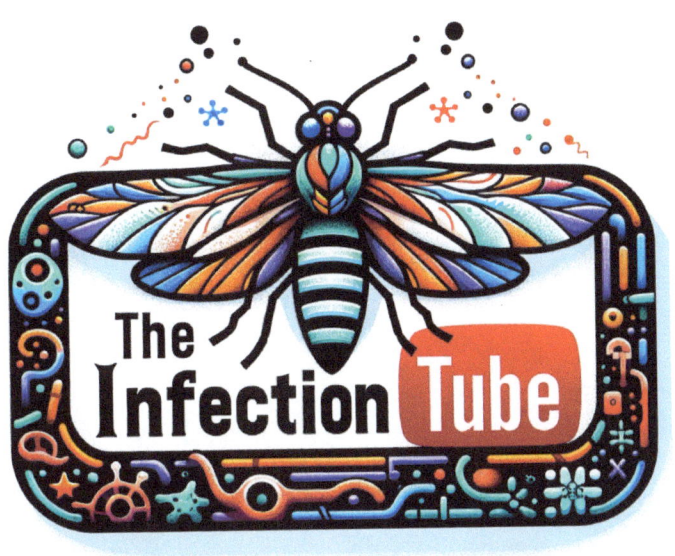

subscripe

THE INFECTION TUBE

Gram negative comedy night

Tonight, in a Microbial Comedy Club where bacteria gather to share laughs, the stage is set for a Gram-negative bacilli stand-up night. This large family has gathered to share funny moments and introduce themselves to each other.

Gram negative comedy night

Here we have **E. coli**, a versatile resident of the human gut,
and
Salmonella, infamous for causing food-borne illnesses.

Gram negative comedy night

Shigella is also here, a key player in bacillary dysentery, alongside

Pseudomonas, notorious for hospital-acquired infections

Gram negative comedy night

In another corner, **Haemophilus influenza,** a common cause of bacterial meningitis in children sits with
Klebsiella pneumonia, known for pneumonia and urinary tract infections.

Gram negative comedy night

And over there, **Legionella pneumophila**, the cause of Legionnaires' disease sits with **Vibrio cholerae**, responsible for the devastating disease cholera.

Gram negative comedy night

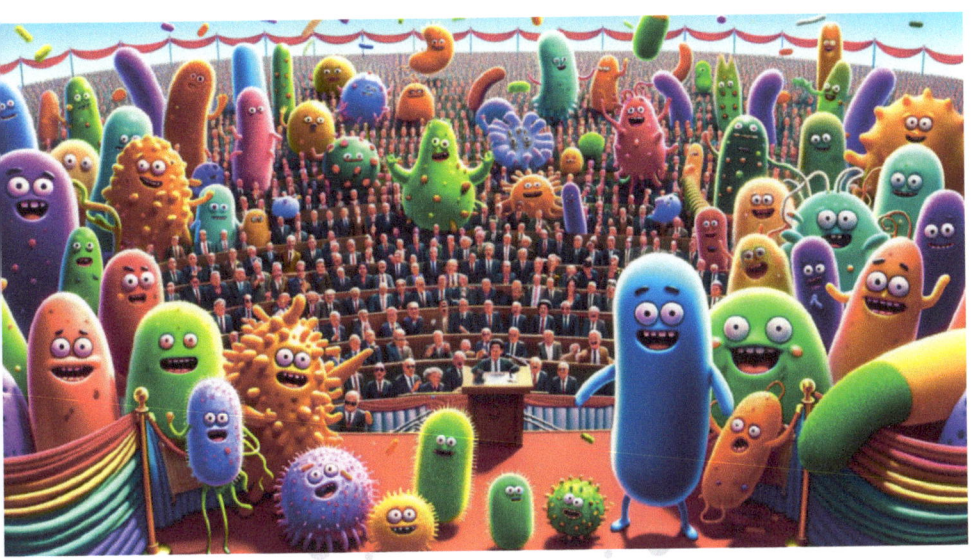

Many other Gram-negative bacilli are present, as well as their cousins, the Gram-positive cocci, viruses, and parasites. What a huge meeting and a funny day it is!!!

Gram negative comedy night

E. coli, sporting a pink cool jacket, steps up to the microphone and said: "Good evening, folks!

I'm E. coli, and unlike my Gram-positive cousins, I don't need a thick coat. My fashion sense is 'thin' in, 'pink' out!

The audience bursts into laughter

Gram negative comedy night

Salmonella said:
And, I'm Salmonella!
I tried cooking once Ended up giving everyone in the room food poisoning

Gram negative comedy night

Pseudomonas aeruginosa, glowing under a black light, says: You think you're cool?
I literally light up a room. Plus,I'm the only one here who can claim to have swum in more pools than Michael Phelps!

Gram negative comedy night

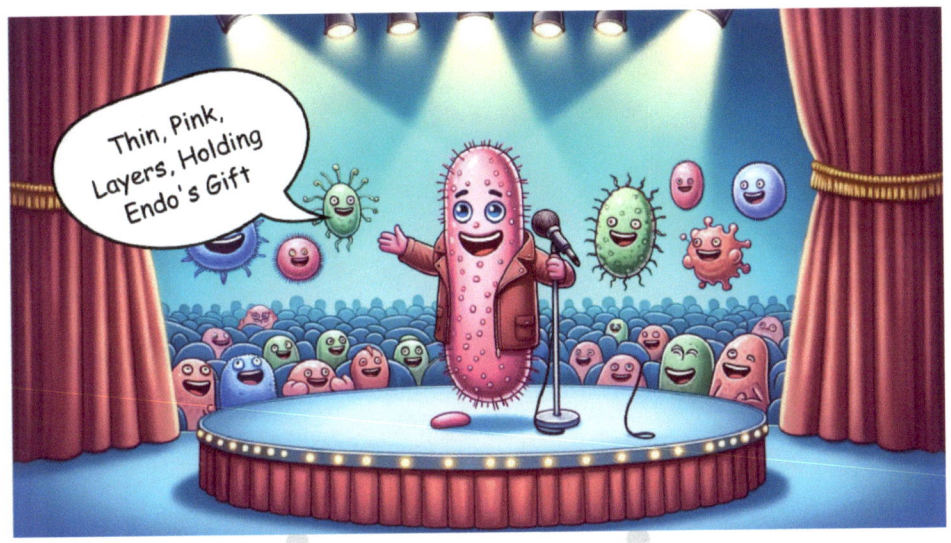

E. coli (returning to the microphone):

But no matter our differences, we all share something special:

'Thin Pink Layers Holding Endo's Gift'

It's not just our structure; it's our legacy

Gram negative comedy night

Salmonella is now getting confused :
"Wait, I thought 'Endo's Gift' was that fruitcake from last year's microbial holiday party.

Gram negative comedy night

Pseudomonas aeruginosa (laughing):"No, that was just another experiment gone wrong. But seriously, 'Endo's Gift' is about our unique lipopolysaccharides.

They might not taste great, but they sure make a lasting impression!

Gram negative comedy night

E. coli concluded:
Remember, folks, in our world, being 'thin' and 'pink' makes you a star!
And our layers? They're not just for looks; they protect us like a knight's Armor

Gram negative comedy night

The three bacilli said together "So, let's hear it for the Gram-negative family!
We might not always be the good guys, but we sure know how to make an entrance and an exit!"

Gram negative comedy night

At the end of the day, the funny talk comes to an end, and the curtain falls on the stage. But you still might be wondering about their own code

'Thin Pink Layers Holding Endo's Gift'

Time for Memory Magic

'Thin Pink Layers Holding Endo's Gift'

'What Does This mean?

Gram negative comedy night

It's a mnemonic in which

"thin" refers to

The thin peptidoglycan layer.

"Pink" represents

Its color after the Gram staining process, as it doesn't turn purple when stained

"layers" referes to

The layered cell wall structure

It's a mnemonic in which

"Holding" **Signifies**

The cell wall's remarkable structural integrity and complexity..... that holds its contents securely

It's a mnemonic in which

"Endo'gift" referes to

The lipopolysaccharides endotoxins that act like a gift, protecting the bacteria from immune defense

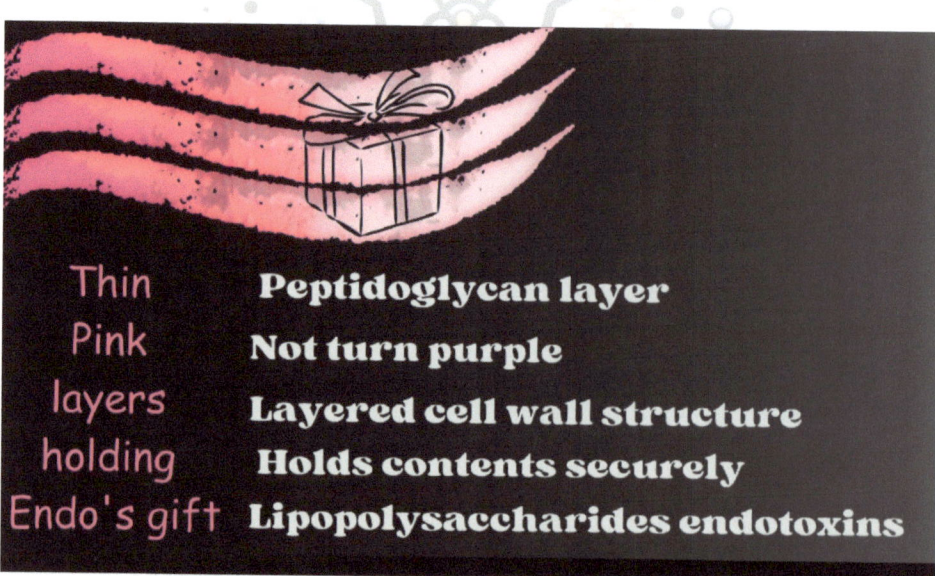

Thin	Peptidoglycan layer
Pink	Not turn purple
layers	Layered cell wall structure
holding	Holds contents securely
Endo's gift	Lipopolysaccharides endotoxins

As the day ends, stay tuned for more microbial adventures! And remember.

knowledge is power!

Continue exploring how your body works and don't forget to subscribe for more insightful videos on The Infection Tube Stay curious and healthy!

Gram negative comedy night

What Knowledge Would You Gain from This eBook?

Common members belongs to the gram-negative family are:

- **Escherichia coli (E. coli)**, a versatile resident of the human gut, with strains ranging from benign to pathogenic.
- **Salmonella**, infamous for causing food borne illnesses.
- **Shigella**, a key player in bacillary dysentery.
- **Pseudomonas aeruginosa**, notorious for hospital-acquired infections.
- **Klebsiella pneumoniae**, known for pneumonia and urinary tract infections.
- **Legionella pneumophila**, the cause of Legionnaires' disease.
- **Vibrio cholerae**, responsible for the devastating disease cholera.
- **Haemophilus influenzae**, once a common cause of bacterial meningitis in children.

Gram negative comedy night

What Knowledge Would You Gain from This eBook?

Understanding Gram-Negative Bacteria:

Structural Characteristics:

- **Peptidoglycan Layer:** Gram-negative bacteria have a thin peptidoglycan layer. Unlike their Gram-positive cousins with thick, robust cell walls, the peptidoglycan layer in Gram-negative bacteria is slender, sandwiched between an inner and an outer membrane. This delicate balance allows them to be both resilient and adaptable.

What Knowledge Would You Gain from This eBook?

Understanding Gram-Negative Bacteria:

Structural Characteristics:

- **Appearance After Gram Staining:** These bacteria appear pink or red after the Gram staining process. This color is a unique badge of honor in the microscopic world, distinguishing them from Gram-positive bacteria, which turn purple. It serves as a visual marker of their family lineage and is a key indicator for microbiologists studying them.

Gram negative comedy night

What Knowledge Would You Gain from This eBook?

Understanding Gram-Negative Bacteria:

Structural Characteristics:

- **Layered Cell Wall Structure:** The cell wall of Gram-negative bacteria is characterized by its layered structure, including a unique outer membrane. This outer membrane offers protection and plays a crucial role in their interactions with the environment. It houses essential components like proteins and lipopolysaccharides, forming a complex fortress that guards against threats.

Gram negative comedy night

What Knowledge Would You Gain from This eBook?

Understanding Gram-Negative Bacteria:

Evolutionary and Functional Insights:

- **Remarkable Structural Integrity:** The intricate design of their cell wall is not just about protection; it is a testament to their evolutionary journey, enabling them to thrive in diverse habitats and conditions. Their cell wall is a shield and a message, showcasing their resilience and sophistication.

What Knowledge Would You Gain from This eBook?

Understanding Gram-Negative Bacteria:

Evolutionary and Functional Insights:

- **Lipopolysaccharides (LPS):** The outer membrane of Gram-negative bacteria contains lipopolysaccharides (LPS), known as endotoxins. This feature is a defining characteristic of the Gram-negative family. LPS can provoke significant immune responses in host organisms, making Gram-negative bacteria formidable players in the microbial ecosystem.

By studying this eBook, you will gain a comprehensive understanding of the unique structural and functional aspects of Gram-negative bacteria, their evolutionary adaptations, and their significant role in the microbial world.

Gram negative comedy night

Test your knowledge

Clinical Scenario 1

Scenario: A 65-year-old man with a history of chronic obstructive pulmonary disease (COPD) presents with fever, productive cough, and difficulty breathing. His sputum culture grows a gram-negative rod that is oxidase positive.

Which organism is the most likely cause of his infection?

- a) Escherichia coli
- b) Salmonella
- c) Pseudomonas aeruginosa
- d) Klebsiella pneumoniae

Answer: c) Pseudomonas aeruginosa
Explanation: Pseudomonas aeruginosa is known for causing hospital-acquired infections, particularly in patients with underlying lung diseases like COPD. It is oxidase positive and a common cause of respiratory infections in these patients.

Gram negative comedy night

Test your knowledge

Clinical Scenario 2

Scenario: A 25-year-old woman presents with severe watery diarrhea after returning from a trip to a developing country. Her stool sample tests positive for a comma-shaped, gram-negative bacterium.

What is the likely causative agent?

- a) Shigella
- b) Salmonella
- c) Vibrio cholerae
- d) Legionella pneumophila

Answer: c) Vibrio cholerae
Explanation: Vibrio cholerae is a gram-negative bacterium that causes cholera, characterized by severe watery diarrhea. It is comma-shaped and commonly contracted in areas with poor sanitation.

Test your knowledge

Clinical Scenario 3

Scenario: A 5-year-old child presents with fever, headache, and a stiff neck. A lumbar puncture reveals gram-negative coccobacilli.

Which organism is the most likely cause of the child's symptoms?

- a) Escherichia coli
- b) Haemophilus influenzae
- c) Klebsiella pneumoniae
- d) Shigella

Answer: b) Haemophilus influenzae
Explanation: Haemophilus influenzae is a common cause of bacterial meningitis in children and appears as gram-negative coccobacilli in cerebrospinal fluid.

Gram negative comedy night

Test your knowledge

Clinical Scenario 4

Scenario: A 40-year-old man with a history of alcoholism presents with a high fever, cough, and rusty-colored sputum. A chest X-ray shows a right upper lobe consolidation.

What is the most likely pathogen?

- a) Escherichia coli
- b) Klebsiella pneumoniae
- c) Pseudomonas aeruginosa
- d) Salmonella

Answer: b) Klebsiella pneumoniae
Explanation: Klebsiella pneumoniae is known for causing pneumonia, especially in patients with underlying conditions like alcoholism. The production of a thick, mucoid, rusty-colored sputum is characteristic of Klebsiella pneumoniae infections.

Gram negative comedy night

Test your knowledge

Clinical Scenario 5

Scenario: A 30-year-old woman presents with dysentery characterized by abdominal pain, tenesmus, and blood in the stool. Stool culture reveals a non-lactose fermenting gram-negative bacillus.

Which pathogen is most likely responsible for her symptoms?
- a) Escherichia coli
- b) Salmonella
- c) Shigella
- d) Pseudomonas aeruginosa

Answer: c) Shigella
Explanation: Shigella is a non-lactose fermenting gram-negative bacillus that causes bacillary dysentery, characterized by abdominal pain, tenesmus, and bloody stools.

Gram negative comedy night

Test your knowledge

Clinical Scenario 6

Scenario: An elderly nursing home resident develops a urinary tract infection (UTI). The urine culture grows a gram-negative rod that is lactose fermenting.

What is the most likely causative organism?
- a) Pseudomonas aeruginosa
- b) Escherichia coli
- c) Shigella
- d) Legionella pneumophila

Answer: b) Escherichia coli
Explanation: Escherichia coli is the most common cause of urinary tract infections and is a lactose-fermenting gram-negative rod.

Gram negative comedy night

Test your knowledge

Clinical Scenario 7

Scenario: A 55-year-old man develops severe pneumonia after returning from a business trip. His urine antigen test is positive for a specific gram-negative rod. **Which pathogen is most likely responsible for his illness?**

- a) Haemophilus influenzae
- b) Klebsiella pneumoniae
- c) Legionella pneumophila
- d) Vibrio cholerae

Answer: c) Legionella pneumophila
Explanation: Legionella pneumophila is known to cause Legionnaires' disease, a severe form of pneumonia, and can be detected using a urine antigen test.

Gram negative comedy night

Test your knowledge

Question 1

Which of the following structures in Gram-negative bacteria contributes to their resilience and adaptability?

- A) Thick peptidoglycan layer
- B) Slender peptidoglycan layer sandwiched between inner and outer membranes
- C) Single membrane
- D) Lack of peptidoglycan layer

Answer: B) Slender peptidoglycan layer sandwiched between inner and outer membranes

Explanation: Gram-negative bacteria have a thin peptidoglycan layer located between an inner and an outer membrane. This structural arrangement contributes to their resilience and adaptability, distinguishing them from Gram-positive bacteria, which have a thick peptidoglycan layer.

Gram negative comedy night

Test your knowledge

Question 2

After the Gram staining process, Gram-negative bacteria appear

- A) Purple
- B) Blue
- C) Pink or red
- D) Green

Answer: C) Pink or red
Explanation: Gram-negative bacteria appear pink or red after the Gram staining process, which is a key visual marker distinguishing them from Gram-positive bacteria that appear purple.

Gram negative comedy night

Test your knowledge

Question 3

Which component of the Gram-negative bacterial cell wall can provoke significant immune responses in host organisms?

- A) Peptidoglycan
- B) Teichoic acid
- C) Lipopolysaccharides (LPS)
- D) Phospholipids

Answer: C) Lipopolysaccharides (LPS)
Explanation: Lipopolysaccharides (LPS), present in the outer membrane of Gram-negative bacteria, are known as endotoxins and can provoke significant immune responses in host organisms.

Gram negative comedy night

Test your knowledge

Question 4

The outer membrane of Gram-negative bacteria houses which of the following essential components?

- A) Teichoic acids
- B) Peptidoglycan
- C) Proteins and lipopolysaccharides
- D) Lipoteichoic acids

Answer: C) Proteins and lipopolysaccharides
Explanation: The outer membrane of Gram-negative bacteria houses essential components such as proteins and lipopolysaccharides, which play crucial roles in protection and interactions with the environment.

Gram negative comedy night

Test your knowledge

Question 5

What is a defining feature of the Gram-negative bacterial cell wall that contributes to their evolutionary success?

- A) Single-layered cell wall
- B) Absence of peptidoglycan
- C) Multilayered cell wall with an outer membrane
- D) Presence of teichoic acids

Answer: C) Multilayered cell wall with an outer membrane

Explanation: The Gram-negative bacterial cell wall is characterized by its multilayered structure, including an outer membrane. This complex design contributes to their structural integrity and evolutionary success, enabling them to thrive in diverse habitats and conditions.

Gram negative comedy night

Test your knowledge

Question 6

Which feature primarily differentiates Gram-negative bacteria from Gram-positive bacteria under the microscope?

- A) The thickness of the peptidoglycan layer
- B) The presence of flagella
- C) The ability to form spores
- D) The shape of the bacteria

Answer: A) The thickness of the peptidoglycan layer
Explanation: Gram-negative bacteria have a thin peptidoglycan layer, while Gram-positive bacteria have a thick peptidoglycan layer. This difference is a primary factor in differentiating the two groups under the microscope after Gram staining.

Gram negative comedy night

Test your knowledge

Question 7

Why is the outer membrane of Gram-negative bacteria crucial for their survival?

- A) It allows for rapid cell division.
- B) It provides structural support and contains endotoxins that can trigger immune responses.
- C) It helps in DNA replication.
- D) It is the site of protein synthesis.

Answer: B) It provides structural support and contains endotoxins that can trigger immune responses.
Explanation: The outer membrane of Gram-negative bacteria provides structural support and houses lipopolysaccharides (LPS), which are endotoxins that can provoke immune responses, playing a critical role in their survival and interaction with host organisms.

Gram negative comedy night

Test your knowledge

Question 8

What role does the thin peptidoglycan layer play in Gram-negative bacteria?

- A) It prevents the cell from drying out.
- B) It serves as a barrier to antibiotics.
- C) It provides structural strength while allowing flexibility between the inner and outer membranes.
- D) It enables the bacteria to move.

Answer: C) It provides structural strength while allowing flexibility between the inner and outer membranes.
Explanation: The thin peptidoglycan layer in Gram-negative bacteria provides necessary structural strength while maintaining flexibility, which is essential for the bacteria's resilience and adaptability.

Gram negative comedy night

Test your knowledge

Question 9

The unique color of Gram-negative bacteria after Gram staining is due to:

- A) The retention of the primary crystal violet stain.
- B) The decolorization step with alcohol or acetone.
- C) The presence of chlorophyll.
- D) The addition of a green counterstain.

Answer: B) The decolorization step with alcohol or acetone.
Explanation: Gram-negative bacteria appear pink or red after Gram staining due to the decolorization step with alcohol or acetone, which removes the primary crystal violet stain, allowing the counterstain (safranin or fuchsine) to be visible.

Test your knowledge

Question 10

Which of the following statements best describes the evolutionary advantage of Gram-negative bacteria's cell wall complexity?

- A) It reduces the need for nutrient uptake.
- B) It enhances their motility.
- C) It provides a multifaceted defense mechanism against environmental threats.
- D) It simplifies their genetic structure.

Answer: C) It provides a multifaceted defense mechanism against environmental threats.
Explanation: The complex cell wall structure of Gram-negative bacteria, including the outer membrane and peptidoglycan layer, provides a multifaceted defense mechanism against environmental threats, contributing to their evolutionary success and ability to thrive in diverse conditions.

Gram negative comedy night

Finally

See the gram negative night in the Microbial comedy club of the infection tube through this link

press Here

Or just scan this QR code